YOUR KNOWLEDGE HAS VALUE

- We will publish your bachelor's and
  master's thesis, essays and papers

- Your own eBook and book -
  sold worldwide in all relevant shops

- Earn money with each sale

Upload your text at www.GRIN.com
and publish for free

Imprint:

Copyright © 2015 GRIN Verlag, Open Publishing GmbH
Print and binding: Books on Demand GmbH, Norderstedt Germany
ISBN: 9783668544000

This book at GRIN:

http://www.grin.com/en/e-book/376863/feasibility-study-of-hybrid-renewable-energy-systems-in-kerala-india

Chirag Anilkumar

Feasibility Study of Hybrid Renewable Energy Systems in Kerala, India

GRIN Publishing

Feasibility Study of Hybrid Renewable Energy System in Kerala

Chirag Anilkumar

Mechanical Engineering Department

Government College of Engineering, Thrissur

Kerala, India

Abstract-- **The study conducts a feasibility analysis on Hybrid Renewable Energy Systems (HRES) in Kerala for sustainable and efficient methods of energy production by tapping the natural resources of wind and photovoltaic solar energy and its application in energy re-production by raising the water released from the penstock after running the turbine for electricity production in micro-hydel projects, back into the reservoir for reuse and re-production of energy using the free energy harnessed through Hybrid Renewable Energy System (HRES) achieved by lifting water through series of height intervals forming small water banks and storage systems at various intervals.**

Index terms-- **Energy Re-production, Hybrid energy system, Renewable energy, Sustainability**

TABLE OF CONTENTS

INTRODUCTION

Energy is a vital input in all sectors of any country's economy. It is crucial for human development index as human development is positively correlated to energy consumption. Till late 1980's energy has been generated largely by burning coal, hydrocarbon oil and natural gas leading to huge carbon emissions. Hence, environmental crisis has become a critical concern for the world today. Emission of greenhouse gases, limited coal availability, environment distortion, rising prices of fossil fuels and pressure on foreign exchange reserves have created hindrance in the prolongation of these resources. Due to this, new energy economy is developing. This new energy economy generates energy from wind, sun and through heat within earth itself. Energy generated by burning fossil fuels damages the environment and causes climate change. However, energy based on renewable sources in general, wind and solar energy specifically, does not affect the environment that adversely, which conventional energy sources do.

Energy crisis is a much debated and threatening problem in Kerala. Therefore an innovative idea needs to be put in place that meets the energy demands of the people and also ensuring a sustainable method for energy production.

A hybrid energy system, or hybrid power, usually consists of two or more renewable energy sources used together to provide increased system efficiency as well as greater balance in energy supply.

The following study conducts feasibility analysis on Hybrid Renewable Energy System (HRES) in Kerala for sustainable and efficient methods of energy production by tapping the natural resources of wind and photovoltaic solar energy.

The knowledge of sustainability and technology is further applied to assess the capability to pump water back into the reservoir of a micro-hydroelectric project through energy derived from a hybrid system of wind and photovoltaic solar energy thereby achieving energy re-production.

NEED FOR HRES

Advantages of HRES include fuel saving (fossil fuel), lower atmospheric contamination, lower maintenance cost, ability to act as a silent system and form an interconnected power system of renewable sources.

Solar radiation received per unit area and wind velocity can vary according to climatic conditions and temperatures.

Hybrid energy network is necessary especially in the nights and rainy seasons, when solar panels are unable to support the constant energy supply to run the motors. Some wind turbines can also be kept for backup in case of sudden climatic changes affecting the light intensity received.

So for a steady and efficient energy output, a hybrid system is mandatory involving both photovoltaic panels and wind turbines.

LIMITATIONS

Limitations include lack of efficient technology and higher costs for storage of the energy harnessed from HRES and energy loss involved during conversion especially for the proposed micro-hydel project. Geographical restrictions for installation of wind turbines and solar panels, high costs of solar panels, climatic changes etc. also add to the drawbacks of HRES.

PROPOSED PLAN

Proposed plan involves analyzing the application of HRES in energy re-production by raising the water released from the penstock after running the turbine for electricity production in micro-hydel projects, back into the reservoir for reuse and re-production of energy using the free energy harnessed through   Hybrid Renewable Energy System (HRES) achieved by lifting water through series of height intervals forming small water banks and storage systems at various intervals.

At various height intervals small water banks should be constructed accordingly for collection of water before pumping them to higher levels. Water banks can be constructed at a level below the tail race coming from the turbines, such that water flows directly into the water banks with the help of gravitational potential set across them. Reciprocating pump can be connected at least 0.5m above the bottom of the tank to prevent unwanted debris such as mud, stones etc from being sucked into the pump.

Later free energy obtained through installation of wind turbines and photovoltaic panels in a hybrid network providing constant energy supply to finally raise the water back to the reservoir for energy re-production.

Subsequent height intervals at which smaller water banks should be constructed for collection of water should be calculated accordingly. Thus, energy obtained from HRES can be used directly without its storage thereby utilizing its maximum potential to proper use.

ADVANTAGES

The free energy harnessed through HRES can be applied directly without the encountering the problems of storage by running the motors that pump water to higher levels at each intervals of height, thereby reducing energy loses during transportation and storage encountered in the earlier systems.

In case of micro-hydel project as proposed, storage of energy produced is uneconomical and unsustainable compared to the payback period calculated.

Even when climatic conditions are unfavourable for suitable energy production, energy can still be normally produced through the present micro-hydel project and water running out can be stored in various water banks till suitable weather conditions are available, for raising the water back to the reservoir.

Solar Radiation

Annual Average: 5.30 (kWh/m^2/day)

Monthly Average

Jan	5.68
Feb	6.24
Mar	6.66
Apr	6.12
May	5.49
Jun	4.04
Jul	4.25
Aug	4.72
Sep	5.36
Oct	4.85
Nov	4.92
Dec	5.22

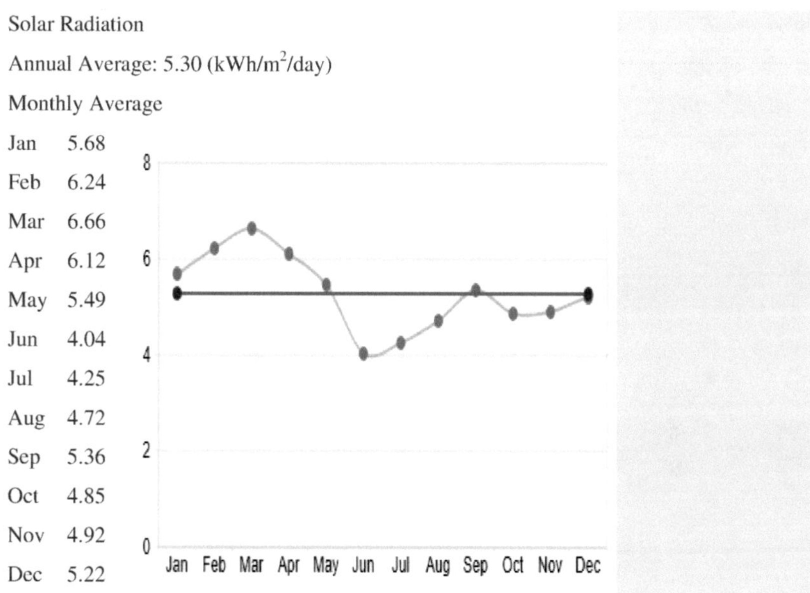

Geographical Information

| Latitude | 10.8505159 |
| Longitude | 76.2710833 |

Wind Speed in Kerala, India (**Source : NREL**)

Wind Speed

Annual Average: 4.30 (mph)

Monthly Average

Jan	3.69
Feb	3.52
Mar	3.74
Apr	3.89
May	4.48
Jun	5.96
Jul	5.59
Aug	5.34
Sep	4.44
Oct	3.57
Nov	3.30
Dec	4.04

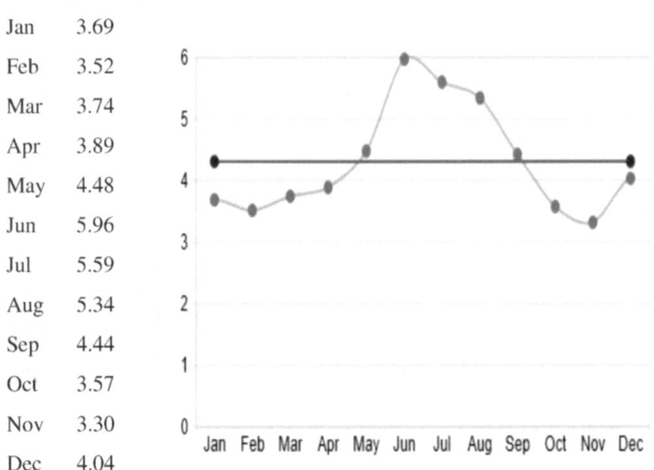

Geographical Information

Latitude	10.8505159
Longitude	76.2710833

POWER GENERATED

From above known and recorded values, considering ideal cases and also accounting for some energy losses involved during the process, an approximate level of output energy can be calculated through following means.

As throughout the year both solar and wind energy vary according to seasons and climatic conditions, a Hybrid system is necessary to obtain considerable power through sustainable means.

For wind turbines, power can be calculated via

$$P = 0.5 \times \rho \times A \times v^3$$

$P = 0.5 \times \rho \times A \times C_p \times v^3 \times N_g \times Nb$ (Equation is neglected for lack of experimental values)

P= Power produced (mechanical) (Watts [W])
ρ=density of wind (kilograms per cubic meter [kg/m^3])
v= velocity of wind (meters per second [m/s^2])

Since wind speed is not constant, a wind farm's annual energy production is never as much as the sum of the generator nameplate ratings multiplied by the total hours in a year. The ratio of actual productivity in a year to this theoretical maximum is called the capacity factor. Typical capacity factors are 15–50%; values at the upper end of the range are achieved in favourable sites and are due to wind turbine design improvements.

According to Betz's law (1919), the wind turbine efficiency of converting the kinetic energy of the wind into mechanical rotating energy can never exceed 59%. Betz's law limits the efficiency of the wind turbine and also the energy output.

Applying the following considerations power generated from wind turbines can be calculated with respect to various months based on the available data.

Symbol	Description	Typical values
ρ	air density	1.2 kg/m^3 (sea level)
Cp	performance coefficient	0.35 is typical 0.56 is the theoretical maximum known as the Betz limit.
Ng	generator efficiency	50 percent to 80 percent.
Nb	Gearbox	95 percent

For Photovoltaic System, power can be calculated via

$$E = A \times r \times H \times PR$$

E = Energy (kWh)

A = Total solar panel Area (m²)

r = solar panel yield (%)

H = Annual average solar radiation on tilted panels (shadings not included)

PR = Performance ratio, coefficient for losses (range between 0.5 and 0.9, default value = 0.75)

For standard test conditions (STC): radiation = 1000 W/m², cell temperature = 25 °C, Wind speed=1 m/s, AM=1.5, the unit of the nominal power of the photovoltaic panel in these conditions is called "Watt-peak".

The solar panel yield of a PV module of 76.615 kWp with an area of 2.5 m² is 91.5%.

Based on above considerations, power generated by solar panels alone can be calculated.

Average Power Generated in proposed study (Theoretical)

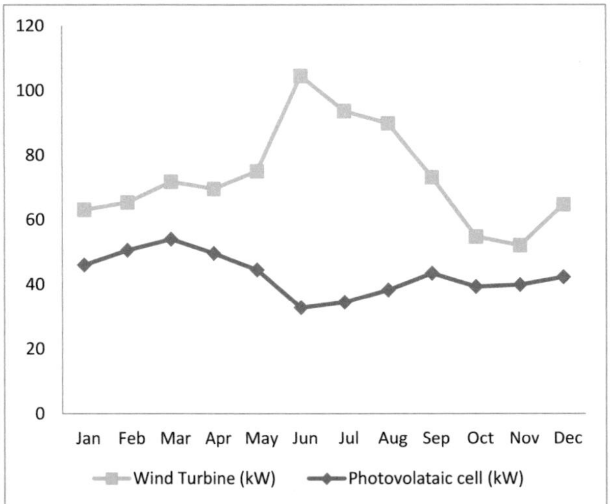

The graph obtained above depicts the theoretical value of power generated with respect to the study undertaken.

Based on the above observations a hybrid system of wind and solar energy systems can be used optimally with respect various months.

Hybrid energy network is necessary especially in the nights and rainy seasons, when solar panels are unable to support the constant energy supply to run the motors.

Wind turbines used successively in series with solar panels can reduce the payback period due greater extent.

FINAL OUTPUT

The quantity of water pumped can be calculated according to the following equation,

$Q = P/\rho \times g \times h$

P = Output power

ρ = Density of water

h = height intervals for pumping water back to the reservoir

Taking 75% of power available to us considering loses, the height considered be each of 25 m for a total height of 100 m and mean power available - 43.377 kW (Wind energy produced can vary from almost 60-100 kW according to fluctuating wind speeds, also solar energy produced can vary according intensity of light received throughout days and months. Therefore, energy received at optimum level involving hybrid network of solar and wind energy will be of higher value)

$$Q = 0.174 \text{ m}^3/\text{s}$$

For such a value we are capable of pumping water back into the reservoir through HRES system to reproduce the energy simply back through turbine system connected to hydroelectric energy system.

PAYBACK PERIOD

Payback period can also be calculated assuming a budget of Rs. 5 crore and the current power rating by the KSEB at Rs 6.5 per watt, as about 15 years.

After completion of almost 15 years, energy produced is profit to the economy and society. Major issue with renewable energy is due to incompatibility in storage and distribution, but HRES involves direct utilization of the energy in all possible means to regenerate power through hydroelectric system, which is easier for service and storage. It also addresses the issues of carbon emissions and polluting atmosphere thereby achieving sustainability at all standards.

RESULT AND ANALYSIS

The above study is suitable for existing micro-hydel systems where solar panels and wind turbines can be installed. Wind turbines can be installed on higher grounds in the existing micro-hydel projects where wind speed is optimum for energy production.

Even when climatic conditions affect both wind turbines and solar panels, energy can still be normally produced from the available micro-hydel project.

Micro-hydel projects are economically viable and environmentally sustainable. It can be constructed at areas where wind speed and solar intensity received is optimum for energy production.

Solar energy can be optimized by the installation   of solar panels of good efficiency to run the motors in pumping water back to the reservoir. Wind turbines limited by geographical and climatic restrictions must be installed after proper study and survey of the climatic conditions of the given area.

Hybrid energy network is necessary especially in the nights and rainy seasons, when solar panels are unable to support the constant energy supply to run the motors.

Some wind turbines can also be kept for backup in case of sudden climatic changes affecting the light intensity received.

Water collected in water banks can also be diverted for agricultural and commercial use, thereby optimizing the usage of water.

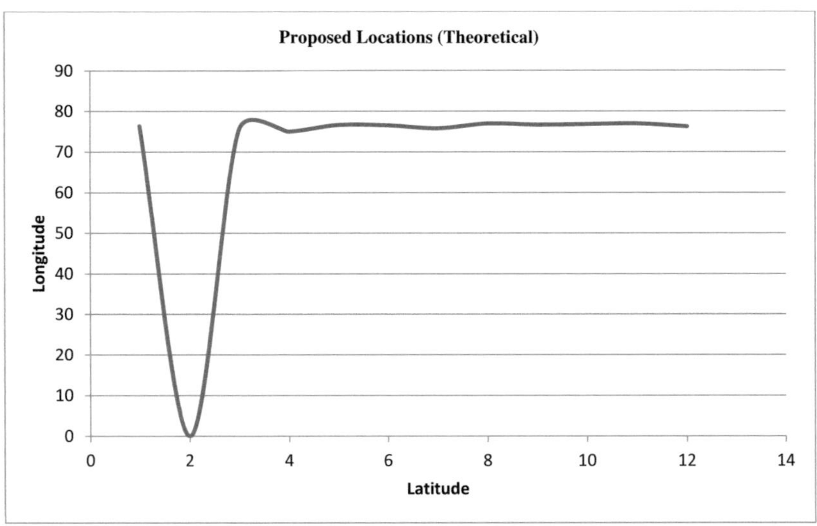

CONCLUSION

According to the data available and by current calculations payback period of 15 years is feasible economically, ensuring sustainability at all standards.

Limitations of HRES can be successfully met by the proposed idea ensuring free energy and sustainability through efficient method of energy re-production for improving the welfare and livelihood for the people.

Almost 30% water resources present in Kerala are now being tapped for energy and agricultural purposes; using innovations such as HRES and its applications we can tap the enormous amount of energy available to us naturally.

Optimizing the potential of solar energy obtained and utilizing wind turbines for replacement and in addition to solar panels, we can further reduce the payback period and help us achieve a sustainable future.

Role of HRES is not only limited to energy generation, but it also contributes to the country by generating employment, reducing adverse effects of greenhouse gases and increasing size of GDP.

ACKNOWLEDGMENT

I thank Aneez Sir, Associate Professor, Government Engineering College Thrissur for supporting and guiding me throughout this study.

REFERENCES

Deepak Sangroya and Dr. Jogendra Kumar Nayak,   "Development of wind energy in India", International Journal of Renewable Energy Research, Vol.5, No.1, 2015

Vanjari Venkata Ramana and Debashisha Jena, "An accurate modeling of photovoltaic system for uniform and non-uniform irradiance", International Journal of Renewable Energy Research, Vol.5, No.1, 2015)

V. Khare, S. Nema, P. Baredar," Status of solar and wind renewable energy in India", Renewable and Sustainable Energy Reviews, Vol. 27, November 2013

A.K. Singh, S. K Parida, "Evaluation of current status and future directions of wind energy in India", Clean Technologies and Environmental Policy, Vol. 15(4), December 2012

Ministry of New and Renewable Energy

National Renewable Energy Laboratory

Yüksel Oğuz, Harun Oğuz, İsmail Yabanova, Emrah Oguz, Ali Kırkbaş, "Efficiency analysis of isolated wind-photovoltaic hybrid power system with battery storage for laboratory general illumination for education purposes " , International Journal of Renewable Energy Research,Vol.2, No.3, 2012)

M. Engin, M. Çolak, "Analysing solar-wind hybrid power generating system", Pamukkale University Faculty of Engineering, Journal of Engineering Sciences, Vol.11, No.2 , 2005.